What's the Worst that Could Happen?

More Bad Assumptions, Ignorance, Failures,

and Screw-ups

in Engineering Projects.

Volume-II

Number 19 in the Computer Architecture Series

Patrick H. Stakem

© March 2018

Table of Contents

Introduction..5

The author..6

Second edition..7

System Failure Case Study's...7

Transportation..9

    Railroad..9

    Self Driving Cars..10

        Tesla Autopilot..11

        Google..12

        Uber and Uber Freight...12

        Ottomotto..12

        Random Access..13

    Submarine disasters..13

        USS Thresher..13

        Russian K-141 Kursk...14

        Submersible Alvin attacked by a swordfish................................15

    Road Tunnel Collapse...15

Personal Electronics...16

    Samsung Galaxy Note 7..16

Aerospace...17

    F22 Raptor, and Y2K..17

    Launch Vehicle Reliability..18

        Atlas Centaur AC-67...18

        ISS and its resupply..18

        US SpaceX Dragon..20

Satellite Payloads..22

    Satellite on-orbit collision..22

    Russian Proton launch Vehicle upper stage..................................22

    Soyuz 23

    Chinese Space Station uncontrolled re-entry................................23

    Snap-10-A..25

    Spaceflight non-fatal accidents...26

Industrial Accidents...26

    Pepcon Disaster...26

    The Honolulu Fireworks Disposal Explosion.................................28

Natural Disasters...28

    Limnic Eruptions..28

    Iceland Volcano shuts down European Air Travel..........................29

- Sinkholes 30
- Geomagnetic Storms 30
- Loma Prieta Earthquake 30
- Meteor Airburst 31
- Near Earth Objects 32
- Medical 32
- Death and injury by Safety Gear 34
- Oil Platform collapse 34

Mean Time to Failure for a Cathedral 35

Extinction Event 36

Afterword 36

Glossary 37

Bibliography 43

- Resources 49

Introduction

"There is no bad situation that you can't make worse" (attributed to Astronaut Corps)

Unfortunately, we learn more from failures than from successes. This book continues an examination of a cross-section of engineering failures, and analyzes them to define the lessons-learned. It also presents some additional methodologies to prevent failures, or, at least, minimize the effects.

The first volume presented a cross-section of failure studies, mostly drawn from the engineering and aerospace context. Each study includes specific references and a definition of the root cause of the failure. I said, "Let's try to learn from other's mistakes. It is less painful to learn from others' failures than your own." So, almost nobody listened, and I have enough material for a new book. We will see continuing errors of omission, errors of commission, and just plain ignorance of the facts. Since its publication, I have had more than enough material for a second book.

The first book discussed System Engineering processes and procedures that can and should be applied to an architecture before the fact, and, unfortunately, after the fact as a post-mortem analysis. It is important to do a good post-mortem analysis of failures, and document them, for the benefit of the next project and the next generation of implementers. This helps to prevent the repeating of mistakes. And yet, we keep repeating mistakes, not building on our work, and going on to create larger, more creative mistakes. I will try to minimize the overlap in the books, but some times there is further information or corrections on material in Volume 1.

A major new area of interest is self-driving cars, not just their technology, but the insurance and legal issues they introduce.

I am presenting this from an engineering standpoint, because that is my background. But the concepts apply across all disciplines and endeavors. If your are responsible for a design, a device, a policy, or a program, you must think through the consequences. Always have a plan B. Always have a Plan C Always think about safety and security. Don't make the same mistake twice. Don't make a second mistake. Keep lessons learned and feed this information forward.

I have included an updated glossary and a list of reference material at the end of the book, and specific references for the failure cases discussed.

Please don't be the clause of a planetary class disaster. Or, at least, please document the case for the benefit of the survivors, if any, to be featured in the next edition of this book.

As I write this, the recent bridge collapse across the highway at the school in Florida has occurred. There were six fatalities. Forensic investigations have begun, but there won't be hard evidence for a few months.

The author

The author spent 42 years in support of NASA spaceflight mission, on this planet and others. He taught for the graduate Computer Science Department of Loyola University in Maryland, Capitol Technology University, Engineering and Computer Science

Department. He teaches for the Johns Hopkins University, Whiting School of Engineering, Engineering for Professionals Program, He has published some 45 technical books. He is somewhat a hands-on expert with major systems failures.

Second edition

The first book was reviewed again after publication, and guess what? More errors were found! How did those get in there? Not quite all of these were corrected for the second edition. I also broadened the scope, and included a lot of new material, including some errors.

Volume 2 of what will be a continuing series on failures presents less theory, and more case studies. Only if failures continue to happen, will this series be a success. Failure, as was said in the first book, is based on complexity. The complexity of a system is a function of the number of parts, and the level of interaction of those parts. A more complex system has more opportunities to fail.

System Failure Case Study's

"Those who cannot learn from history are doomed to repeat it," George Santayana.

This section discusses selected case studies of systems that went wrong. Strangely, I don't have to look very hard for examples. There is a wealth of data available.

Forensic Engineering is the discipline that looks into the determination of causes, after the disaster has occurred. This data, hopefully, will be including in engineering best practices. It also provides data for court cases related to the incident. Early forensic investigations were made of 19th century bridge collapses (We're still not getting that right).

In one case in 1847, a train fell through a bridge that had been designed by famed engineer Robert Stephenson. It went to inquest, and was determined to be Stephenson's fault.

Let's talk about the iconic Liberty Bell, in Philadelphia. It was cast in England, and sent to the Colonies. It cracked the first time it was wrung. It was melted down and recast locally, with added copper content. It then sounded really bad. It was re-melted and silver was added. That resulted in a good tone. In 1835, for the funeral of Chief Justice John Marshall, it was rung again, and cracks appeared. It stays silent now.

Maybe the Brits aren't good at bells. The iconic Big Ben in the House of Parliament in London. showed cracks, so it was equipped with a smaller clapper, oriented differently. The Chiming mechanism suffered catastrophic failure in 1976 when the shaft of the fly governor failed. There was much damage in the clock room, but its working again. A manufacturing flaw was discovered in the shaft, but it had lasted for 4 million striking cycles.

There is a systematic way to identify risks, and to reduce or mitigate them. This is referred to as Disaster Risk Reduction. It is based on research, and past case studies. The UN Office for Disaster Risk Reductions says DDR is, "The conceptual framework of elements considered with the possibilities to minimize vulnerabilities and disaster risks throughout a society, to avoid (prevention) or to limit (mitigation and preparedness) the adverse impacts of hazards, within the broad context of sustainable development." The UN addresses large humanitarian disaster, but their defined approach reaches across all instances and disciplines. (https://www.unisdr.org/)

Transportation

"History repeats itself because no one was listening the first time." Anonymous.

This section discusses failures in earthly modes of transportation, aircraft, highway, water, and rail. The (U. S.) National Transportation Safety Board (NTSB) keeps extensive databases of aviation events at: http://www.ntsb.gov/investigations/reports_aviation.html

from their Mission Statement:

"The National Transportation Safety Board is an independent Federal agency charged by Congress with investigating every civil aviation accident the United States and significant accidents in other modes of transportation – railroad, highway, marine and pipeline. The NTSB determines the probable cause of the accidents and issues safety recommendations aimed at preventing future accidents."

Railroad

It is getting more and more difficult to find a documented steam locomotive incident. The latest is from 1995, and concerns a boiler explosion on the Gettysburg Railroad. The locomotive had completed two excursion runs that day, and the crew were getting ready for a third. It appears that they fireman had let the boiler water level get too low, and the engineer did not notice. On a grade, the crown sheet of the boiler became uncovered. Thermal stress caused the sheet to fail, and the boiler exploded into the firebox. Superheated steam escaped through the firebox door into the cab, injuring three. Before the explosion, the steam pressure had dropped to 175 psi, from 230 psi. The engineer and firemen were trained for their positions, but not specifically on a steam locomotive. The water gauge and gauge cocks (valves) had not been tested before the trip, nor were they "blown down," a procedure that makes sure they are not clogged. For some reason, the feedwater pump gauge had been removed. This is necessary to see that the pump is overcoming boiler pressure to continue adding water. Maybe a little more on-the-job training was needed.

Self Driving Cars

One of the predecessor technology's to self-driving cars is lane assist, or lane departure warning. This story was related to me by one of my students in the Embedded Systems class at JHU. "My dads car has lane-assist built into it. I'm assuming the device is a small camera that simply monitors the color patterns of the road, and sends an error signal to the ECU when it determine the vehicle is getting to close to the yellow line. The ECU will handle the rest, normally trying to correct the position of the car. One time we were driving somewhere and he had forgotten to turn the feature off, and we entered a construction zone, where the traffic had to be redirected across some lines. Trying to drive the car in the new traffic pattern, the car was registering the older paint lines and attempted to correct the car. Unfortunately it was trying to correct the car into a cement wall, although luckily my dad was paying attention. Systems like this seem to be programmed for ideal road situations, and don't take into account PennDOT doesn't paint over the old lines. Similar situations have occurred with Tesla's auto-pilot systems. A big fear I have of these systems, that even if a fail-safe is built in, such as it turning the system off when it is reading confusing road markers, is that people will get so used to it being there that they will assume the car will take care of it."

Cars are getting safer and more intelligent. There's still a long ways to go. They know all the rules of the road, see 360 degrees, and they are better at driving than most humans, but there are still issues to be resolved.

The U.S. Highway Traffic Safety Organization (NHTSA) defines automotive automation in 5 levels. Level 0 is no automation, which includes your Studebaker Pick-up truck. Most 2018 cars implement Level 1, which includes lane departure and antilock brakes. Level 2 is like Tesla's Autopilot, which can take over from the driver when it thinks it should. There are situations (one described below) where the unit needs human intervention. If it gets overwhelmed with events, it asks for help. Level 3 is limited self-driving automation. Level 4 is where a human need not touch the controls.

Tesla Autopilot

The Tesla autopilot feature is responsible for the first death in a driverless car. There are some bugs to be worked out in the technology, and there are attitudes and approaches concerning the car's abilities that need changing.

Joshua Brown's Tesla drove into the side of a white tractor-trailer in full sunlight at 70 miles per hour, proving fatal to Joshua and the car. The car had alerted him to the danger. Even with the autopilot on, the passenger is supposed to keep their hands on the wheel. He didn't have his hands on the wheel (or his attention on the road) for 37 of the 41 minute journey. Why bother, the car drives itself, right? On the car's side, the sensors at the front bumper probably saw underneath the trailer.

The National Transportation Safety Board issued a 500 page report on the incident, after a large amount of research.

Google

Google's approach is completely opposite of Tesla's. Google's car does not have manual controls; it is the driver, and you are the passenger. A Google Lexus SUV drove into the side of a bus. There were no injuries to humans. Google has run simulations, and blame the incident on a lack of negotiation between drivers, a common assumption among drivers, that the electronics doesn't understand. The hardest task for a self-driving car is understanding human behavior.

Uber and Uber Freight

Uber is addressing Robot Taxi's and self-driving big rigs. The field tests for the Taxi's took place in Pittsburgh, with the first accident taking place in Tempe, Arizona. The Uber Volvo wound up lying on its side, after a driver making a left turn failed to yield. There was a human in the Volvo at the time of the accident. The Volvo had entered the intersection on a yellow light, and was held responsible.

Just as I was ready to send this book to press, Uber managed to have a fatality, the first involving a self-driving car, and a pedestrian. This happened in Tempe, Arizona, which is trying to be self-driving car friendly. A back-up driver was in the car. The car was an Uber-modified Volvo XC90, operating in autonomous mode. It was in a collision with Elaine Herzberg, walking her bicycle in the street. The car was traveling about 40 mph, and, apparently, did not see her. A NTSB Team is assisting local authorities. I guess the tombstone inscription will say," killed by faulty software."

Ottomotto

This U. S. company, with veterans of the Google car effort, is developing self-driving automated big-rig trucks. The system can be retrofitted on existing units. There are currently three Volvo big-rigs in test. Another player in this field is Peloton Technology, who are addressing "platooning' where a convey of trucks move together. The advantages are that the trucks can operate in closer formation than those with human drivers.

Random Access

This is relevant to self-driving cars, but did not involve one. Yet. In this scenario, hackers took over a Jeep, and operated it remotely. The driver was not happy. He was doing 70 MPH on a highway near St. Louis, when he noticed his air conditioning came on by itself. Then, the radio came on at full volume. The wipers came on. A picture came up on the dash display, of the two hackers. This was not actually a surprise to the driver, it was a scheduled test. Using the entertainment system in the vehicle, the steering, brakes, and transmission could be commanded from the hacker's location, 10 miles away. This was a controlled experiment, before a planned presentation at the Black Hat Security Conference. The hackers shared the details of the exploit with the car manufacturer.

Submarine disasters

This section discusses a few undersea problems. When you have a problem in a submarine, you have a big problem.

USS Thresher

The Thresher, (SSN-593) sunk in 1963 during deep driving trials a few hundred miles east of Boston. All 129 crew were killed. This lead the Navy to implement a new submarine safety program. The Thresher was the fastest and quietest submarine in the world at its launch. In Navy tradition, the nuclear-powered Thresher was not decommissioned, but remains on Eternal Patrol.

The boat was built at the Portsmouth Naval Shipyard, launched in 1960, and commissioned in 1961. She conducted extensive tests, returned to the shipyard for checkout, and then returned to sea floor diving trails. She was accompanied by the submarine ship Skylark. They lost communications with the sub above but near her test depth. Fifteen ships went to the search area. By morning the next day, the hunt was abandoned.

Shattered remains were later located on the ocean floor at a depth of 2600 meters. The Trieste deep diver was deployed, and managed to photograph the wreckage.

A Navy Court of Inquiry concluded the Thresher had probably suffered the failure of a salt-water piping system joint which relied heavily on **silver brazing** instead of welding.

The site is monitored for ambient radiation, and none so far has been found. The Navy had Robert Ballard attempt to photograph the sub, after which it bankrolled his Titanic Expedition.

A sister-sub, the Scorpion, was the only other American nuclear submarine to be lost as sea. One of the theories of her sinking involves a malfunction of a trash disposal unit.

Russian K-141 Kursk

The nuclear-powered Kursk was lost in the Barents Sea in the year 2000, with 188 personnel on board. This followed an attempt to launch a series of practice torpedoes. These were powered by hydrogen peroxide. It was postulated that a faulty weld allowed this to leak, ignited the kerosene fuel, and destroyed the torpedo room. The fire caused the detonation of multiple torpedo warheads. The boat had recently won a citation for its excellent performance and been recognized as having the best submarine crew in the Northern Fleet.

All personnel onboard were killed. There was evidence that 23 sailors survived the initial explosion, but died 6 hours later in another compartment when an oxygen generator caught fire and used up the remaining oxygen.

Kursk was was found seven days after the accident. The hull was recovered by a Dutch Company under contract. All but the bow was discovered, and all bodies were found. The published report alluded to "stunning breaches of discipline, shoddy, obsolete and poorly maintained equipment," and "negligence, incompetence, and mismanagement."

Submersible Alvin attacked by a swordfish

Alvin is a submersible of the Woods Hole Oceanographic Institution, and has been making exploration dives since 1964.

This event happened in 1967, when a crew was onboard, at a depth of 2,000. The perpetrator was stuck on the sub, and had to be removed, after an emergency surfacing. In retaliation, it was then eaten. Seriously, I can't make stuff like this up.

Road Tunnel Collapse

Whether you are driving, or the car is driving itself, you don't want the tunnel you're in to collapse. Unfortunately, this is what happened in Boston, in 2006, to the East bound tunnel to the Airport. Roughly 12 tons of concrete panels had let loose and crashed down into a car. There was one fatality. Luckily, the traffic volume in the tunnel was low. The determined cause was anchor bolts letting loose from the ceiling, as the epoxy to hold them failed. It was found there was a lack of understanding on the project of epoxy. Epoxy had not been identified as a safety-critical component. It was shown that the adhesive had undergone creep deformation and fracture. The epoxy used was known to have poor creep resistance under load.

"People should not have to drive through the Turnpike tunnels with their fingers crossed." Mitt Romney, Governor, Massachusetts

The NTSB learned during the investigation that Fast Set epoxy had been tested for creep performance in 1995 and 1996 (by an independent testing laboratory) and failed to meet the standard. There is a required 120 test protocol, and the epoxy failed this test. In addition, the number of bolts used had been reduced. The NTSB cited MTA (Massachusetts Transit Authority) for failing to implement an aggressive inspection program in the aftermath of known roofing system failures in 1999 and 2001.The repairs took 2 years, and over $50 million.

Personal Electronics

This section discusses problems with our personal electronics that we rely on. For example, in 2014 the *Nest Fire Detector* was recalled due to a feature that allowed the alarm to keep from sounding. The product was recalled preemptively, before any injury or incidents were reported. Over 440,000 detectors were recalled. You sometimes don't want to hear those pesky smoke alarms, I guess.

Samsung Galaxy Note 7

This premium cellphone had a nasty habit of catching fire, The company had to recall in excess of 2.5 million units. They were banned from aircraft. Some newer models, like the S7, had this issue as well. The company denies responsibility, and has cited external damage to some of the failed units. These models were recalled and discontinued, and the company's operating profits suffered a big hit. Some caught on fire spontaneously in user's pockets.

Caught on camera, a Dell laptop caught on fire on a couch in February 2017 while charging. Luckily, some one was home to unplug it and extinguish the flames, or the house would have been lost. A teenager in Canada was using a third party charger for his i-phone, but it caught on fire and ignited the bed.

Aerospace

Getting spacecraft off the surface of the planet and into orbit, or to other planets is somewhat difficult. There are many documented failure cases in this realm, not because the rocket scientists and engineers make more mistakes, but due to the difficulty of the endeavor. This is an area of particular interest to me, as I worked as a NASA support contractor since the 1970's on many missions. I have a lot of case studies in this area. A much more exhaustive study in the area is the book by Harland and Lorenz, dated 2005, which is in the reference section at the end of this book. There is a lot more material that happened since 2005, unfortunately. To date, 18 astronauts have been killed in in-flight accidents. Early German rocket propulsion pioneer Max Valier died in an explosion in 1930.

F22 Raptor, and Y2K

Built by Lockheed Martin, the F-22 Raptor is an advanced stealth U. S. fighter jet. It entered service in 2005, and now has mostly been replaced by the F-35.

In February 2007, on the aircraft's first overseas deployment from Hawaii to Japan, a 15 hour flight, six F-22s of the 27th Fighter Squadron flying from Hawaii, experienced multiple software-related system failures while crossing the International Date Line. They lost navigation, communications, and some other onboard systems. Luckily, they were refueling at the time, and the refueling planes stayed with them, escorting them back to Hawaii. Ooops. The aircraft landed ok, and the software error error was fixed in a couple of days.

Launch Vehicle Reliability

The first tricky part is getting the payload off the ground. This has been shown to be a hazardous task since the initial work by Von Braun in Germany, and Robert Goddard, in the U. S. Let's keep in mind that Robert Goddard was supposedly asked to stop his rocket launches by the Fire Marshall. If the launch vehicle fails, the payload is irrelevant. We'll look at a few launch vehicle failures.

Atlas Centaur AC-67

Lightning from thunderstorms can be catastrophic to a rocket if it strikes in mid flight. In 1987, lightning struck a Atlas Centaur AC-67 rocket less than a minute after liftoff, causing it to explode.

ISS and its resupply

The first part of the International Space Station went into orbit in 1998. It has been expanded since then, to such an extent that it is visible to the naked eye. Most of the construction and "heavy lifting" was done by the United States and Russia, with support from the European and Japanese Space Agencies. The ISS is actually the ninth inhabited Space Station in Earth Orbit, with previous Russian (USSR) and U. S. efforts. It has been continuously occupied, as of this writing, for more than 15 years. It has rotating crews from the United States and Russia, with other member countries supplying personnel, and paying customers – space tourists. It is periodically reboosted to higher altitudes, as its orbit decays due to atmospheric drag.

It has been necessary to occasionally conduct Debris Avoidance Maneuvers, to avoid know space debris nearing the station. It has been hit many times, but the hull has not been breached.

Although numerous failures and problems have affected Station operations and schedule, no one has been hurt of killed, and it has never been evacuated.

There are periodic resupply and logistics flights, carrying repair parts, new experiments, clean underwear, food, oxygen, and water. Return flight carry trash, and burn up in the atmosphere. Some small payloads can be returned with the returning crew.

The Space shuttle was the primary supply "truck" for the Station, both during its construction and for part of its operational period. The Shuttle fleet was retired in 2011.

The Soyuz capsule is used to take crews up and back, but has limited cargo space. The Russian *Progress 62* vehicle is used for upgoing cargo It has a capacity of 2.8 tons.

Some of these flights have had problems. M-12 failed to reach orbit in 2011. Around 325 seconds into the flight, the third stage of the Russian Soyuz-U rocket prematurely shut down, leaving Progress M-12M stranded on a sub-orbital trajectory. The failure, the first ever for a Progress since its introduction in 1978, could not have come at a worse time for the ISS, with the Space Shuttle recently retired, and commercial resupply flights not yet online.

The Progress 59 mission experienced serious problems shortly after its launch on April 28, 2015, and never attempted to dock with the station. Then, SpaceX's seventh cargo run failed less than three minutes after launch on June 28, when the Falcon 9 broke apart, apparently because of a faulty steel strut in the rocket's upper stage.

US SpaceX Dragon

In 2012, Dragon became the first commercial spacecraft to deliver cargo to the International Space Station. SpaceX delivered cargo to the ISS in March 2013 and again in April 2014. The Dragon vehicle is designed to return to the surface, as opposed to the other units, which burn upon reentry. The Space-X capsule is recovered at sea. In February of 2015, the Dragon capsule returned 3,700 pounds to Earth from the station.

In June of 2015, a SpaceX resupply mission to the ISS failed when CRS-7 failed to reach orbit. This was attributed to a booster problem. The Company Press Release said, "Preliminary analysis suggests the overpressure event in the upper stage liquid oxygen tank was initiated by a flawed piece of support hardware (a "strut") inside the second stage. Several hundred struts fly on every Falcon 9 vehicle, with a cumulative flight history of several thousand. The strut that we believe failed was designed and material certified to handle 10,000 lbs of force, but failed at 2,000 lbs, a five-fold difference."

In order to reduce launch costs, Elon Musk's private space company SpaceX is making a reusable booster from its existing Falcon-9 launch vehicle. The unit will have enough propellant to land vertically in a selected location, after delivering its payload to orbit. This will allow for refurbishment and reuse of the booster, which provides a major cost savings.

The first several flights were landed in the ocean, to show proof of concept. The first flight suffered excessive roll rates during descent. The fuel was flung to the outside walls of the tank due to centrifugal forces, and the engines flamed out. Some debris was recovered. The second test achieved a soft landing, but the vehicle was not recovered due to rough seas. The third and fourth flights achieved zero velocity at zero altitude, but the boosters fell into the ocean (as planned). The next flight was scheduled to land on a floating recovery ship. Rough seas in the recovery area, with waves up to 3 stories, were a major problem. Adding to that, one of the thrusters that kept the barge in position failed. The launch vehicle landed in the ocean.

The next attempt resulted in a hard landing on the target barge, and the vehicle fell overboard, and sank. One the next attempt, there was another hard but successful landing, that damaged some of the barge's deck equipment. The seventh landing attempt was also hard, due to a stuck valve, and the stage tipped over on the barge, and burned.

An attempt to land the returning booster on dry land was accomplished successfully in December of 2015. More tests will be conducted.

The Falcon booster costs $60 million, and the cost to orbit will be vastly reduced with reuse of the expensive hardware. These cases are not the worst, they are just learning experiences with new technology approaches.

The Orbital STK Cygnus vehicle uses Orbitals' Antares launch vehicle, and has a capacity of 7,000 pounds. It has suffered 3 failures in 4 launch attempts. In October 2014, the launch vehicle exploded just after leaving the launch pad.

Satellite Payloads

Even if the payload survives the trip to orbit on the launch vehicle, it can still have problems. We'll explore several case studies here. If it is hard to get to orbit, it is very hard to get to another planet, and increasingly difficult to land there.

Satellite on-orbit collision

A collision between two satellites occurred in February of 2009. One was a Russian Strela-class military satellite, massing 950 kilograms. The other was the commercial Iridium 33 communications satellite. What was the cause? They were in the same place at the same time. The Russian spacecraft had been deactivated, and was classified as space debris. The Iridium was operational, and was destroyed.

And, the bad news is, the collision created a thousand pieces of space debris larger than 4 inches, and many more smaller ones. In March 2012, a piece of the KOSMOS 2251 passed by the International Space Station, prompting the crew to take refuge in the attached Soyuz return craft as a precaution. The ISS frequently does obstacle-avoidance maneuvers.

Russian Proton launch Vehicle upper stage

The importance of designing for proper passivization has been demonstrated by the more than 40 ullage motors flown on the Russian Proton Block DM upper stage that have broken up in orbit. The ullage motors, first deployed in the 1980s, provide the stage with three-axis control during coast, and are routinely ejected when the Block DM stage ignites for the final time. Depending on the mission profile, the ullage motors may carry up to 40 kilograms of unused propellant. Over time, solar heating and other factors have caused dozens of the motors to explode, releasing debris into orbit. Russia has made design changes to prevent accidental explosion of the engines.

Soyuz

The former Soviet Union suffered two disasters, and one near-disaster, all three involving the capsule during the de-orbit and re-entry. Soyuz-1 ended in disaster when the parachutes failed to deploy and the capsule smashed into the earth at speeds over 300 mph, killing cosmonaut Vladimir Komarov. Soyuz 5 almost ended in disaster, when the re-entry capsule entered the atmosphere by the wrong end – attributed to a failure of the equipment module to separate similar to that on the Vostok-1 flight. Luckily, the equipment module burned off the descent module and the capsule righted itself.

Soyuz 11 ended in disaster in 1971 when an equalization valve, used to equalize air pressure during the Soyuz final descent, prematurely opened in the vacuum of space, killing the three crew members, who were not wearing spacesuits. Subsequent flights, from Soyuz-12 to Soyuz-40, utilized a two-man crew because the third seat had to be removed for the pressure suit controls. The Soyuz-T version restored the third seat.

Chinese Space Station uncontrolled re-entry

As this book was being written, we are waiting to see where the out-of-control Chinese Tiangong-1 Space Station is going to hit. It is massive enough to cause extensive damage if it goes anywhere except the satellite graveyard in the South Pacific.

The point in the Oceans farthest from land is called the pole of inaccessibility. This is the best point to aim for, if you're de-orbiting something large. It happens to be located in the South Pacific, some 1,600 miles south of the Pitcairn Islands. There are some 260 satellites on the ocean floor at that point. This will be an interesting place for future archaeologists. Among other things, the 120-ton MIR space station is there. It's entry into the water was observed by fishermen. This is also where supply modules from the ISS, loaded with trash, are sent. Sometimes, things don't quite work out. 36 tons of Salyut Space Station came down on land in South America.

Launched in 2011, the 9.4 ton station ceased functioning in March 2016. It was put in sleep mode in 2013. It is the largest spacecraft object to-date to fall uncontrolled.

The liability for space debris rests with the United Nations Convention on International Liability for Damage Caused By Space Objects. The nation (or nations) that launched the object is "absolutely liable" for compensating the injured party, according to the convention.

Just one person, Lottie Williams of Tulsa, Oklahoma, has ever been hit by space junk, according to news reports, but she wasn't hurt. On Jan. 22, 1997, Williams saw "a flash of light resembling a meteor." A few moments later, something metallic fell onto her shoulder. NASA reportedly said her incident came close to the timing of the re-entry and breakup of the second stage of a Delta rocket coming into Earth's atmosphere.

Snap-10-A

I could have put this in one of two places, and I chose this arbitrarily. This project was a really bad idea, not having any thought given to the consequences.

The SNAP, or Systems for Nuclear Auxiliary Power, was a satellite launched to LEO in 1965. It stopped working after 43 days in orbit, due to an electrical problem, and the core automatically shut down. It will remain in that orbit for an estimated 4,000 years. It was noticed in November of 1979 that the satellite was shedding debris, possibly due to a collision.

There are actually more than 30 small nuclear reactors in orbit, mostly Soviet. To validate the safety systems on SNAP-10A, a destructive experiment was conducted in 1964, leading to the spread of radioactive material across a desert in Idaho.

The Demonstration of Autonomous Rendezvous Technology (DART) mission was a 2001 program was to rendezvous with a target satellite, the Multiple Paths, Beyond Line-of-sight Communications (Mublcom). It managed to collide with the target. This was shown to be caused by inaccurate measurement of speed and distance. This, in turn, was found to be caused by an infinite loop in the navigation system, causing resets. This was blamed on lack of software validation and verification. Misuse of heritage software and design flaws were also blamed. The target satellite was not damaged, and continued to operate. The DART used up all of its maneuvering propellant, and was considered a mission failure.

Spaceflight non-fatal accidents

There are many more non-fatal accidents than fatal ones, with equipment loss and personnel injury. Before the Apollo landings were attempted, astronauts trained in the Lunar Landing Research and Training Vehicles. Three of the five crashed and were destroyed, Number 1 crashed at Ellington Air Force Base in Texas, due to a loss of helium pressure, leading to loss of the steering jets. Luckily, Neil Armstrong ejected safely.

Industrial Accidents

This section discusses a few Industrial accidents with wide ranging effects.

Pepcon Disaster

This explosion occurred in Nevada in 1988 at the Pacific Engineering and Production Company of Nevada (PEPCON) plant. It produced ammonium percolate, an oxidizer used with solid fueled rocket engines, such as those of the Space Shuttle. There were two killed and 372 injuries. The damage was estimated at over $100 million. The cause is unknown. Interestingly, there are only two producers of the chemical in the U. S., with the other plant being 1.5 miles away.

When the disaster occurred, there were some 4500 metric tons of the product in storage at the plant. There were numerous secondary explosions, and first responders turned back after getting close to he site, and staged some 1.5 miles away. A five mile radius around the plant was evacuated.

The explosions leveled the plant, and a near-by marshmallow manufacturing facility. The shock wave from the major explosion knocked a car off the highway several miles away, killing the driver.

The exact cause of the explosion is unknown. A lot of the evidence was destroyed, and investigators had to rely on eyewitness reports, and video. Based on damage, the explosion was estimated to be the equivalent of about 40 kilotons off TNT.

Imperial Sugar Company Explosion

You usually don't think of sugar as an explosive, but in an aerosol form, it is. This was demonstrated in 2008 at the Imperial Sugar Company Refinery in Georgia. After the main explosion, fires burned for seven days. Fourteen workers were killed, and many others injured. Actually, any material as fine dust particles can be explosive. Grain elevators used to explode before this was fully understood. OSHA says, any particles with a diameter of 420 microns or less are explosive in confined spaces. Additionally, the explosion's shock waves frees up more collected dust, causing a chain reaction.

The facility had been built in 1917, and specialized in converting raw cane sugar into granulated sugar. It was producing 2.6 billion (10^9) pounds of sugar per year, at the time of the incident.

Some of the plant machinery was undersized and out-dated. Workers said that there was a persistent cloud of sugar dust, and it was inches to feet deep on the floors. The ignition source was probably sparks from an electric motor.

The Honolulu Fireworks Disposal Explosion

In an attempt to make things safer, the Island of Oahu set up a storage and disposal area to deal with imported fireworks seized by federal authorities. There were no specific safety standards in place. In the end, 5 workers were killed.

The State of Hawaii, Dept. of Health, issued a emergency hazardous waste permit, in 2010, allowing the fireworks to be properly disposed of. They were collected at a WW-II Navy munitions storage bunker. A lot of the black powder was removed from the devices and stored in plastic garbage bags.

Military Explosive Ordnance Disposal had assisted in the disposal of fireworks, until three technicians had been killed in this activity. The job was handed over to "the low bidder." The actual cause of the explosion was not determined, but was thought to have come from a spark from a metal tool, or static discharge from a plastic bag, igniting material that had accumulated on the floor of the facility. There were new regulations related to adequate safety training of personnel.

Natural Disasters

This section discusses a range of natural disasters that we have to deal with.

Limnic Eruptions

In a limnic eruption, dissolved carbon dioxide erupts from deep water lakes and forms a cloud that can suffocate living organisms, including humans. There can also be a resulting tsunamis, causing flooding. The locations are known as exploding lakes. They are not necessarily related to volcanic activity.

Mass Causality events occurred in the 1980's at Lake Monoun and Lake Nyos, in Cameroon. The first instance resulted in 38 deaths, and the second in 1,700 deaths. These eruptions are rare, and not completely understood.

Iceland Volcano shuts down European Air Travel

In 2010, the eruption of the Eyjafjallajökull volcano shut down air traffic across Europe due to the spreading dust cloud. If in-jested into the jet engines, it will clog them, and they will fail. In 1982, a British Airways Flight flew through volcanic ash from Mount Galunggung in Java, causing all 4 engines to shut down. The crew was able to restart all but one and landed safely. This triggered an addition to the operations manuals, describing how to deal with volcanic ash

This was the largest air travel shut-down since World War-II. The eruption was beneath glacial ice. This caused the lava to quickly cool and break into very small particles. Glacial meltwater flowed into the eruption vent causing ash to be injected up into the jet stream. It was estimated that the airline industry loss was $200 million per day, total loss around $1.7 billion ($10^9$). The closure of the airspace left some five million travelers stranded. Some military aircraft were still flying, and significant engine damage was noted.

The London Volcanic Ash Advisory Centre provided information updates to the various air carriers. In excess of 95,000 flights were canceled. This event caused aircraft manufacturers to publish new guidelines on ash injection, which allowed aircraft to fly when there are levels of volcanic ash between 200 and 2000 micrograms of ash per cubic meter.

I went to Iceland to see that Volcano. There a very nice interpretive center, built and operated by a family who lived through the eruption, near-by. Definitely a must-see. The volcano is still steaming. What's the worst that could happen?

Sinkholes

Sinkholes that form suddenly are of interest.. You shouldn't be standing there when they form. Also occurring, but not related, are ground collapses, due to human activity, such as underground mining. In the United States, Florida has the most sinkhole incidents, sometimes involving entire houses and neighborhoods disappearing. Sink holes can also form under the sea. The Bayou Corne sinkhole in Louisiana spans some 25 acres.

Geomagnetic Storms

The March 1989 severe geomagnetic storm lacked out Quebec for some 9 hours. It was caused by a solar coronal mass ejection, that hit the Earth 3 days later. The charged particles induced electric currents in the ground. On-orbit satellites were affected. Currents were induced in very long electrical transmission lines, such as those of Quebec's power grid. Radio communications were also adversely affected.

Loma Prieta Earthquake

The Loma Prieta earthquake struck on October 17, 1989, at 5:04 p.m. It was a 6.9 event. There were 63 deaths, and almost 4,000 injuries. Game3 of the World Series was scheduled to start at 5:35 p.m. at Candlestick Park and thousands of people were already in the stadium when the quake occurred. There was a delay of game. People around the country were going to watch that game (including me), but it was the first time an earthquake event had been captured live, and broadcast.

The 3rd game of the 1989 World Series. The series involved the San Francisco Giants and the Oakland Athletics, both local teams, so the stadium was packed. An estimated $6 billion (109) was caused by the earthquake. The Oakland-San Francisco Bay Bridge was heavily damaged. A section of freeway took 11 years to repair.

Broadcaster Al Michaels said on the air, "I'll tell you what, I think we're having an earth..." There were 62,000 fan in the park at the time. Both Team's stadiums were damaged, but play resumed in 10 days.

Meteor Airburst

A meteor entering the Earth's atmosphere can cause an explosion. These may have started out as asteroids in space. The extreme example is the 1908 Tunguska Event in Siberia, an explosion equal to an atomic weapon. In 2013, the Chelyabinsk meteor event, also in Russia, was widely photographed, and went viral on the internet. The meteor size was estimated to be 20 meters.

Asteroids of up to 4 meters in size impact the Earth about once per year, with 3 kilotons of energy. Asteroids of up to 70 meters impact every 1900 years or so, with an energy of 16 megatons (TNT equivalent). These can trigger extinction events.

There was a documented case in 1490 in China, *the Ch'ing-yang event*, that resulted in 10,000 deaths. A bolide is an extremely bright meteor, that usually explodes in the atmosphere. It can also refer to a crater-forming event.

The Earth Impact Effects Program characterizes the average energy of airbursts, and the average frequency of the event. This ranges from a 4 meter, 0.7 kiloton airbusrt with an average frequency of an every every 1.4 years, to a 70 meter, 15 megaton event, every 1900 years.

We know more about meteor airbursts now, with increased monitoring under the Comprehensive Nuclear-Test-Band treaty, an improved detection technology. In 2017, there were 26 events noted.

Near Earth Objects

Technically, an NEO is a solar system object whose closest approach to the Sun is 1.3 AU, and that comes in close proximity to the Earth. There are 14,000 known asteroids in this category, 100 comets, solar orbiting spacecraft, and meteoroids. All these have the potential of striking the Earth. They are closely tracked from the ground, by NASA's Planetary Defense Coordination Office. A joint US/EU project called Spaceguard is tracking NEO's larger than 30 meters. Three NEO's have been visited by spacecraft.

Medical

(this is courtesy of a student in my Embedded Systems class, at JHU.)

A leading manufacturer of ventilation products for the Hospital and pre-hospital markets with and annual revenue of 9.6 Billion Euros, invests 8% of that revenue into the R&D efforts. In 2017 the company was forced to issue a recall on some of its products due to a software problem. The recall affected all of the gold standard hospital dedicated noninvasive ventilator with version

2.20 software release. The software defect could erroneously report the blower motor has stalled and cause the unit to shut down leading to hypoxemiaor hypercarbia (hypercapnia) due to lack of oxygen. Despite the company's effort to make this product safe, the FDA declared this failure to be a Class I recall (the highest-risk label on a recall) that affects more then 19,000 units.

The fault is falsely triggered but the software does behave appropriately to the failure it wrongly detected; once detected the device will displaying an "E100 error code", and shut down while alarming in a high priority alarm mode. The company is a leader in the ventilator space and this recall was determined to "produce a situation in which there is a reasonable probability that the use of or exposure to the product will cause serious adverse health consequences or death." Coincidentally, there was a recall a year earlier for potential motor failures that affected 116 units and was related to an issue in the blowers ventilator assembly.

Evidently, pacemaker software is not known for its reliability but is known for its vulnerability to hacking. A recent study defined 8,000 known vulnerabilities in the code. A very small percentage of manufacturers address security, making their devices wide open to cyber attack. There is also supposed to be a login code and password, for when the device needs an upgrade. Even this was ignored. The U.S. Food & Drug Administration has jurisdiction over companies that make medical devices. Less than half of the manufacturers were found to be following advise from the FDA on device security.

Death and injury by Safety Gear

Safety gear and Personal Protective Equipment is meant to step in when we encounter a dangerous situation. Sometimes it doesn't work like that.

Failure of PPE, root causes.

- Incorrectly worn PPE, or not fitted or used correctly
- PPE that is not cared for or stored correctly
- Incompatible PPE – Items of PPE which (worn simultaneously) undermine each other's effectiveness
- Inappropriate PPE – Wearing PPE that is not fit for purpose
- Uncertified PPE – Untested, Unproven, Unreliable

Oil Platform collapse

In 2001, the largest offshore deepwater oil platform in the world collapsed. Not too big to fail. This was 150 km off the coast of Brazil, and the platform was owned by the state energy company Petrobras. There had been a series of explosions that killed 11, and collapsed one of the main support columns. The platform had been built in 1994.

Emergency drain tanks, used to store excess fluids that were trapped or under excessive pressure are normally installed under the bottom desk. This tower's tanks were installed in the two rear support columns, to save money. The main driver for the project was reducing costs, not safety. About 1300 cubic meters were released from the tanks. Of course, the feed water line to the fire fighting pumps were damaged. The fire fighters were in the wrong place at the wrong time, and 10 were killed.

Five days later, the platform sunk in 1,350 meters of water, as a total loss.

Mean Time to Failure for a Cathedral

The Gothic style Beauvais Cathedral in 1284 France suffered a major collapse of its vaulting in the choir. This started a new thinking about cathedrals, and introduced a atmosphere of fear and conservatism among architects. The collapse may have been caused by wind gusts. There was little understanding of buildings and structural mechanics since the Greek and Romans. The Cathedrals were meant to be vast inside, with your eyes drawn up to the heavens. Unfortunately, doing this in stone had reached its limits, even with exterior buttresses. Stone is good in compression, but a bad building material in tension – it can literally be pulled apart. It was not until the advent of iron structures that tension members could be used. In some cases, a large chain was wrapped around the top of the building to take the tension forces. Today, the repairs on Beauvais are not yet complete, and laser scanners and 3-D models are being used to analyze the structure.

The National Cathedral and the Washington Monument in Washington, D. C. both suffered damage due to a rare earthquake in the area in 2011.

We should also mention the Leaning Tower or Pisa. Some one underestimated the weight, and the softness of the ground. It actually started to lean during construction in 1173. Probably built by the low bidder. Major restoration work took place in 1990, reducing the angle of the lean from 5.5 to just under 4 degrees. It should be good to go for another 300 years.

Extinction Event

An extinction event is defined as "a widespread and rapid decrease in the amount of life on Earth." Given that over 98% of the species we know of are extinct, we need to be careful with the home planet. There have been several mass extinctions on Earth since life began, none because of us. So far.

Afterword

Well, the answer is, there's a lot that can go wrong, even in the simplest cases. Systems are designed by humans, and humans are fallible. We can reduce the probably of failure, but we can't completely eliminate it. To err is human, they say, to really screw things up requires a computer.

There was a lot of material to choose from in pulling this book together. I didn't want to go to extremes, so I left out a lot of detail of well know cases: the eruption of the volcano Krakatoa, the "Year without a summer," the Big Northeast Blackout, and many more. You don't need to look far to find similar stories. There's a lot more material relating to incidents that occurred since this book came out. Are we getting any smarter, and more ready? Sleep well.

If you're now curled up in an underground bunker in the fetal position, you missed the point of the book. We can influence systems, we can plan for contingencies, we can learn the lessons of the past and apply them.

If you really want to see the worst that can happen, see the movie, "The Martian."

Glossary

Ada – a software language

AFB – Air Force base.

AI – artificial intelligence.

APU – auxiliary power unit.

Bends – a painful condition, resulting from the nitrogen dissolved in the blood stream collecting in the joints,

BLS – (U.S.) Bureau of Labor Statistics

Blue Screen of Death – an error page in the Windows operating System

Buran – the Russian space shuttle.

CDC – (U. S.) Centers for Disease Control

CME – Coronal Mass Ejection on the Sun

Cntrl-Alt-Del – the three key combination used to reset a pc; developed by IBM to intentionally be difficult to do accidentally; sometimes called the three-fingered salute.

CNSA – China National Space Administration.

CPSC – U. S. Consumer Product Safety Commission.

CPU – central processing unit (computer)

Cyber-terrorism – terrorist acts committed over the Internet.

Deflagration – exothermic reaction that propagates at less than the speed of sound.

DOSH – (U. S.) Division of Occupational Safety and Health.

DRC – Disaster Research Center, U. Delaware.

DRR – Disaster Risk Reduction

Duplicate – provide two solutions.

ECU – engine control unit.

EIEP – Earth Impact Effects Program.

Error segmentation – keeping the effects of an error from cascading.

EU – European Union

EVA – extra-vehicular activity; going outside the spacecraft.

FAA – (U.S.) Federal Aviation Administration.

Fail-safe – a failure does not cause a fault.

Fault Tolerant – a property of a system where 1 or more faults won't cause it to fail.

Fault Tree – a graphical representation of faults and causes.

FDA – (U. S.) Food & Drug Administration.

FEMA – (U. S.) Federal Emergency Management Agency,

Firewall – a computer that connects a local computer to a wider network, and provides protection.

Flash – a non-volatile memory, like in flash-drives.

Flight Software – software used onboard space missions.

Floating point – a computer number format, like scientific notation.

FMEA – Failure Modes and Effects Analysis.

GAO – (U. S.) General Accounting Office.

HFE – Human Factors Engineering

Hubris – from the Greek, extreme pride or self-confidence; sometimes found in engineering design.

Hypergolic - components spontaneously ignite when they come into contact with each other.

IRBM – intermediate range ballistic missile.

ISS – International Space Station.

IV&V – independent verification and validation – having some one else look over the system.

LEO – Low Earth orbit.

MIR – Russian space station; since reentered the atmosphere.

LAN – local area network.

LLIS – (NASA) lessons learned information system

LLRV – lunar lander research vehicle (for Apollo)

LOCV – loss of crew and vehicle.

LTAP – lateral turn across path

Malware – malicious software

micron – micrometer, 10^{-6} meter.

MTBF – Mean time between failures

MTTF – Mean Time to failure.

MTTR – Mean time to Repair.

Mutex – a software mechanism to provide mutual exclusion in processes.

NASA – the U.S. Space agency

NEA – Near Earth asteroid

NEO – Near Earth object.

NFPA – U. S. National Fire Protection Association.

NTSB – (US) National Transportation Safety Board – responsible for accident investigations.

NSF – (U.S.) National Science foundation

Pitot tube – device on an aircraft measuring speed via air pressure differences.

Plan B – what to resort to when Plan A fails.

Post mortem – investigation after the fact.

PPE – personal protective equipment.

PRA – Probabilistic Risk Assessment.

Priority Inversion – a condition in a real-time operating system, in which a high priority task is held up by a low priority task holding a key resource. Demonstrated on the surface of Mars.

PSI – pounds per square inch, a measure of pressure.

RCA – Root Cause Analysis

Redundant – providing multiple units, either completely identical; or functionally identical.

Relay – electromagnetic switch

Reset – return to a known, initial condition.

RFNA – red fuming nitric acid, an oxidizer.

RoHS – restriction of hazardous substances.

Root cause - the first cause, the starting point of events.

Roscosmos – the Russian space agency

RPN – risk priority number, in a FMEA.

Rush-hour – inappropriately named morning and evening event when traffic moves at a crawl.

SCC – stress corrosion cracking.

SE&I – Systems Engineering and Integration.

SI – System Internationale (metric)

Soyuz – Russian manned space capsule, still in use.

SRAM – static random access memory.

SRB – (Shuttle) solid rocket boosters

SSME – Space Shuttle Main Engine (liquid propellant).

Stall – condition in which there is not enough air moving over the wings of an aircraft, and it looses lift.

STS – Space Transportation System (Shuttle).

Third rail (3rd rail) – a power distribution system for subways.

Tram – a light rail system used for passenger service in cities. A bus on rails.

Triplicate – provide three solutions.

TPS – (Space shuttle) Thermal protection system – tiles.

UDMH - Unsymmetrical dimethylhydrazine, a rocket fuel. Hypergolic with nitrogen tetroxide.

USFA – United States Fire Administration.

UXO – unexploded ordnance.

V2V – vehicle to vehicle (communications)

VAERS – CDC's Vaccine Adverse Event Reporting System

Voting logic – choose the majority, based on the assumption that 2 failures are less probable than one.

Zombie-Sat – out-of-control non-responsive satellite posing a danger to other spacecraft.

Bibliography

Berk, Joseph *Systems Failure Analysis*, ASM International, November 17, 2009, ISBN-1615030123.

Bible, George *Train Wreck: The Forensics of Rail Disasters*, Johns Hopkins University Press , 2012, ASIN: B0099SKJXW.

Bozzano, Marco and Villafiorita, Adolfo *Design and Safety Assessment of Critical Systems*, Auerbach Publications; 1st edition, November 12, 2010, ISBN-1439803315.

Burgess, Colin; Kate Doolan, Kate; Vis, Bert F*allen Astronauts: Heroes Who Died Reaching for the Moon*, 2003, Bison Books, ISBN-0803262124.

Burrough, Bryan *Dragonfly: NASA and the Crisis Aboard Mir,* 1998, 1st ed, HarperCollins, ISBN-0887307833.

Burrough, Bryan *Dragonfly: The Terrifying Story of Mir, Earth's First Outpost in Space,* 1999, fourth Estate Ltd. ISBN-1841150886.

Burrough, Bryan *Dragonfly: An Epic Adventure of Survival in Outer Space,* Harper Perennial; reprint edition, march 1, 2000, ASIN-B010EW4GMQ.

Charette, Robert N. "Air France Flight 447 Crash Causes in Part Point to Automation Paradox," July 10, 2012, http://spectrum.ieee.org.

Cheng, P.G. *100 Questions for Technical Review,* Aerospace Report No. TOR-2005(8617)-4204. Space and Missile Systems Center. September 30, 2005.

Dunn William R. *Practical Design of Safety-Critical Computer Systems*, July 2002, ISBN-0971752702.

Durant, Will and Ariel, *The Lessons of History*, 1st ed, 1968, Simon & Schuster, ISBN- 143914995X.

Eliot, Dr. Lance B. *Self-Driving Cars: The Mother of All AI Projects": Practical Advances in Artificial Intelligence (AI)*, 2017, ISBN-0692914544.

Fawcett, Bill *100 Mistakes that Changed History: Backfires and Blunders That Collapsed Empires, Crashed Economies, and Altered the Course of Our World,* Berkley, 2010, ASIN B0042JSOSA.

Fawcett, Bill *Trust Me, I Know What I'm Doing: 100 More Mistakes That Lost Elections, Ended Empires, and Made the World What It Is Today,* Berkley, 2013, ISBN-0425257363.

Fowler, Kim *Mission-Critical and Safety-Critical Systems Handbook: Design and Development for Embedded Applications* Newnes; 1st edition, November 20, 2009, ISBN- 0750685670.

Gill, Paul S. ; Garcia, Danny *Engineering Lessons Learned and Systems Engineering Applications,* NASA, avail:

https://www.researchgate.net/publication/242185667_Engineering _Lessons_Learned_and_Systems_Engineering_Applications

Kardon, Joshua B. *Guidelines for Forensic Engineering Practice*, 2nd ed, 2012, ISBN- 0784412464.

Lipson, Hod; Kurman, Melba *Driverless: Intelligent Cars and the Road Ahead* (MIT Press), 2017, ISBN-0262534479.

https://www.researchgate.net/publication/242185667_Engineering_Lessons_Learned_and_Systems_Engineering_Applications

Harland, David M. and Lorentz, Ralph D. *Space Systems Failures, Disasters and Rescues of Satellites, Rockets and Space Probes*, Springer, 2005, ISBN 0-387-21519-0.

Hermann, Debra S. *Software Safety and Reliability: Techniques, Approaches, and Standards of Key Industrial Sectors*, Wiley-IEEE Computer Society Press; 1st edition, February 10, 2000, ISBN0769502997.

Hobbs, Chris *Embedded Software Development for Safety Critical Systems*, Auerbach Publications, 2015, ISBN-1498726704.

Jones, Capers *Patterns of Software System Failure and Success*, International Thomson Computer Press (December 1995), ISBN-10: 1850328048.

Kichenside, Geoffrey *Great Train Disasters: The World's Worst Railway Accidents*, 1997, Parragon Plus, ISBN-10: 0752522299.

Kieffer, Susan W. *The Dynamics of Disaster*, W. W. Norton & Company, 2013, ASIN: B007Q6XLHK.

Kalinsky, David "Architecture of safety-critical systems, 2005, http://www.embedded.com/design/prototyping-and-development/4006464/Architecture-of-safety-critical-systems#

Klotz, Irene "Programming Error Doomed Russian Mars Probe," Feb. 7, 2012, avail: http://news.discovery.com/space/

Krämer, Bernd J. and Völker, Norbert (Eds.) *Safety-Critical Real-Time Systems*, December 3, 2010, ISBN-1441950192.

Krantz, Gene (2001). *Failure Is Not an Option: Mission Control from Mercury to Apollo 13 and Beyond,* New York: Simon & Shuster. ISBN 978-0-7432-0079-0.

Leveson, Nancy G. "Software Safety in Embedded Computer Systems," Communications of the ACM. Vol. 34, No. 2, February 1991. pp. 34-46.

Leveson, Nancy G. *System Safety and Computers*, Addison-Wesley, 1995, ISBN 0-201-11972-2.

Leveson, Nancy G. *Engineering a Safer World: Systems Thinking Applied to Safety*, The MIT Press, January 13, 2012, ISBN-0262016621.

Linenger, Jerry M. *Letters from Mir,* 2003, McGraw-Hill, ISBN-0-07-140009-5.

Lochbaum, David; Edwin Lyman, Edwin; Stranahan, Susan Q. *Fukushima: The Story of a Nuclear Disaster,* The New Press, 2014, ASIN: B00EXCAJKC, ISBN: 1595589082.

Noon, Randall K. *Introduction to Forensic Engineering*, 1992, CRC Press, ISBN-0849381029.

Noon, Randall K. *Forensic Engineering Investigation,* 2000, CRC Press, ISBN-0849309115.

Offit, Dr. Paul A. *Pandora's Lab: Seven Stories of Science Gone Wrong*, National Geographic, 2017, ISBN-1426217986

Petroski, Henry *To Forgive Design: Understanding Failure*, Belknap Press of Harvard University Press (March 30, 2012), ISBN-10: 0674065840.

Petroski, Henry *To Engineer Is Human: The Role of Failure in Successful Design*, Vintage, 1992, ISBN-10: 0679734163.

Petroski, Henry *Success through Failure: The Paradox of Design* Vintage, 1992, ISBN-10: 0679734163.

Ross. Steven S. *Construction Disasters: Design Failures, Causes and Prevention* (Engineering News-Record Series), McGraw-Hill, 1984, ISBN-0070538654.

Ryan, R. S. "A History of Aerospace Problems, Their Solutions, Their Lessons," NASA Technical Paper 3653, 1996,

 avail:https://ntrs.nasa.gov/archive/nasa/casi.ntrs.nasa.gov/19970001339.pdf

Schlager, Neil (Ed) *When Technology Fails: Significant Technological Disasters, Accidents, and Failures of the Twentieth Century*, Gale Research (1994), ISBN-10: 0810389088.

Schoete, Brandon; Sivak, Michael *A PRELIMINARY ANALYSIS OF REAL-WORLD CRASHES INVOLVING SELF-DRIVING VEHICLES*, 2015, UMTRI-2015-34, U. Michigan, Transportation Research Institute, http://umich.edu/~umtriswt/PDF/UMTRI-2015-34.pdf

Sgobba, Tommaso; Rongier, Isabelle *Space Safety is No Accident: The 7th IAASS Conference*, 2015, ISBN-331915981X.

Spark, Nick T. *A History of Murphy's Law,"* Periscope Film, 2006, ISBN 0-9786388-9-1.

Storey, Neil *Safety-Critical Computer Systems*, Addison-Wesley, 1996. ISBN: 0-201-42787-7.

Sweet, Justin; Schneier, Marc M. *Legal Aspects of Architecture, Engineering and the Construction Process*, 9th Edition, 2012, ISBN-1111578710.

Theodossopoulos, Dimitris; Sinha, Braj "Structural safety and failure modes in Gothic vaulting systems," 8th International Seminar on Structural Masonry, 2008, Instanbul.

Tolker-Nielsen, Toni "Exomars 2016 – Schiaparelli Anomaly Inquiry," 2017, ESA, DG-I/2017/546/TTN.

Turtledove, Harry, *Supervolcano: Eruption* (Novel), Roc; Reprint edition, 2011, Penguin Group (USA) LLC, ISBN: 0451464206, ASIN B005ERIKF6.

Vaughan, Diane *The Challenger Launch Decision: Risky Technology, Culture, and Deviance at NASA*, 2016, ISBN-10: 022634682X.

Vaughan, William, et al "Engineering Lessons Learned and Systems Engineering Applications," 2005, avail: https://www.researchgate.net/publication/242185667

Vogel, David A. *Medical Device Software Verification, Validation and Compliance* Artech House; November 30, 2010, ISBN-1596934220.

Wallis, L. A., Greaves, I *Injuries Associated with Airbag Deployment, Emerg Med J.* 2002;19:490-493 doi:10.1136/emj.19.6.490.

Wichmann, Brian A. *Software in Safety Related Systems*, Wiley, 1992. ISBN: 0471-93474-7.

Resources

"Air Traffic Control Computer Failures: Hearings before a subcommittee of the Committee on Government Operations, House of Representatives, second session, June 30 and August 15, 1980," University of Michigan Library 1980, ASIN: B00300GEMK.

NASA Better Mechanisms Needed for Sharing Lessons Learned, January 2002, United States Government Accounting Office, Report to the Subcommittee on Space and Aeronautics, committee on Science, House of Representatives, GAO-02-195.

Wikipedia (www.wikipedia.org), various.

Useful websites

National Transportation Safety Board, https://www/ntsb.gov

Consumer Product Safety Commission, www.cpsc.gov

National Fire Protection Association, www.nfpa.org

http://www.space.com/10694-human-spaceflight-dangers-infographic.html

Levenson, Nancy sunnyday.mit.edu/accidents/

http://www.embedded.com/columns/technicalinsights/169600396.

This is the website for a new series on the Science Channel that has some fairly well research and well presented space-realated disasters:

http://www.sciencechannel.com/tv-shows/secret-space-escapes/

http://www.iflscience.com/5-most-hair-raising-moments-history-spaceflight

http://www.spacesafetymagazine.com/space-disasters/

http://spectrum.ieee.org/cars-that-think/transportation/self-driving/fatal-tesla-autopilot-crash-reminds-us-that-robots-arent-perfect

https://www.usatoday.com/story/news/nation-now/2017/03/25/ubers-self-driving-car-involved-arizona-crash/99619462/

http://www.theverge.com/2016/2/29/11134344/google-self-driving-car-crash-report

https://www.bls.gov/iif/tables.htm   BLS – Injuries, Illnesses, and Fatalities

http://rebrn.com/re/til-that-on-july-the-alvin-submersible-was-attacked-by-a-swordfi-319074/

http://www.films.com/id/9417/When_Engineering_Fails.htm

CBS, 60 minutes, Anderson Cooper, May 1, 2016, "Strikethrough: A Matter of Life, Death, and Litigation"

http://spectrum.ieee.org/cars-that-think/transportation/self-driving/fatal-tesla-autopilot-crash-reminds-us-that-robots-arent-perfect

https://www.usatoday.com/story/news/nation-now/2017/03/25/ubers-self-driving-car-involved-arizona-crash/99619462/

http://www.theverge.com/2016/2/29/11134344/google-self-driving-car-crash-report

https://nsc.nasa.gov/SFCS/

http://www.dailytech.com/Lockheeds+F22+Raptor+Gets+Zapped+by+International+Date+Line/article6225.htm

http://www.theverge.com/2014/5/21/5739326/440000-nest-protect-smoke-detectors-recalled

https://www.wired.com/2015/07/hackers-remotely-kill-jeep-highway/

http://rebrn.com/re/til-that-on-july-the-alvin-submersible-was-attacked-by-a-swordfi-319074/

http://www.naval-technology.com/features/featureperil-in-the-depths---the-worlds-worst-submarine-disasters-4191027/

http://www.films.com/id/9417/When_Engineering_Fails.htm (51 minute video)

NOVA - "Building the Great Cathedrals." www.pbs.org/wgbh/nova/ancient/building-gothic-cathedrals.html

http://www.dailytech.com/Lockheeds+F22+Raptor+Gets+Zapped+by+International+Date+Line/article6225.htm

"F-22 Squadron Shot Down by the International Date Line." Defense Industry Daily, 1 March 2007. Retrieved: 5 February 2014.

http://standards.nasa.gov

NASA Lessons Learned Information System – http://llis.nasa.gov

PPE - http://blog.prochoice.com.au/ppe/ppe-failure-could-you-be-at-risk/

https://www.osha.gov/SLTC/personalprotectiveequipment/

http://infectioncontrol.tips/2016/05/05/personal-protective-equipment-strikethrough-risk-rhetoric/

National Academy of Forensic Engineers http://www.nafe.org/

"Giant Oil Rig Sinks Off Brazilian Coast," *Environment News Service*, March 20, 2001.

avail: www.ens-newswire.com/ens/mar2001/2001-03-20-01.html

https://www.researchgate.net/publication/242185667_Engineering_Lessons_Learned_and_Systems_Engineering_Applications

Meteor airbursts

- Rubtsov, Vladimir; Ashpole, Edward *The Tunguska Mystery* (Astronomers' Universe) Copernicus; 2009 ed, 2010, Amazon Digital Services, Inc. ASIN: B008BAGYTC.

- Marcos, C.; Marcos R. *The Chelyabinsk Superbolide: we didn't see that one coming,* The Baetylus Press; 1stt ed, 2013, Amazon Digital Services, Inc. ASIN: B00GWWCKK8 .

- NASA's Planetary Defense Co-ord office. https://www.nasa.gov/planetarydefense

- Friend, Tad *Planet Killers*, Byliner Originals, 2011, ISBN 978-1-61452-007-8.

- https://www.newscientist.com/article/mg12517013-200-technology-sorry-no-numbers-the-day-the-uss-telephone-network-crashed/

- https://www.clarkfountain.com/blog/2017/february/5-reasons-your-airbag-might-not-deploy/

If you enjoyed this book, you might find something else from the author interesting as well. Available on Amazon Kindle.

Stakem, Patrick H. *What's the Worst That Could Happen? Bad Assumptions, Ignorance, Failures and Screw-ups in Engineering Projects*, 2014, PRRB Publishing, ASIN-B00JSH540.

Stakem, Patrick H. *Extreme Environment Embedded Systems*, PRRB Publishing, ASIN-B01AF9CBM0, Jan, 2016.

Stakem, Patrick H. *Cubesat Engineering*, PRRB Publishing, 2017, ASIN- B01N4VC99B.

Stakem, Patrick H. Robots and Telerobots in Space Applications, 2011, PRRB Publishing, ASIN B0057IMJRM.

Stakem, Patrick H. *Earth Rovers: for Exploration and Environmental Monitoring,* 2014, PRRB Publishing, ASIN B00MBKZCBE.

Stakem, Patrick H. The History of Spacecraft Computers from the V-2 to the Space Station, 2013, PRRB Publishing, ISBN-1520216181.

Stakem, Patrick H. The Saturn Rocket and the Pegasus Missions, 1965, 2013, PRRB Publishing, ASIN-B00BVA79ZW.

Stakem, Patrick H. Spacecraft Control Center, 2015, PRRB Publishing, ASIN-B01D1Y5LZ0.

Stakem, Patrick H. Embedded Computer Systems for Space, 2015, PRRB Publishing, ASIN-B018BAYCCM.

Stakem, Patrick H. Microprocessors in Space, 2011, PRRB Publishing, ASIN-B0057PFJQI.

Stakem, Patrick H. Apollo's Computers, 2014, PRRB Publishing, ASIN-B00LDT217.

Stakem, Patrick H. Crewed Spacecraft, 2017, PRRB Publishing, ISBN-1549992406.

Stakem, Patrick H. Crewed Space Stations, 2017, PRRB Publishing, ISBN-1549992228.

Stakem, Patrick H. Rocketplanes to Space, 2017, PRRB Publishing, ISBN-1549992589.

Stakem, Patrick H. Visiting the NASA Centers, 2017, PRRB Publishing, ISBN-154965120X.

Stakem, Patrick H. Deep Space Gateways, , 2017, PRRB Publishing, ASIN-B077XJZQ1Y .

Stakem, Patrick H. Manufacturing in Space, 2017, PRRB Publishing, ASIN-B078KJ2RVQ,

Stakem, Patrick H. NASA's Ships and Planes, 2017, PRRB Publishing, ASIN-B078SH4P82.

Stakem, Patrick H. Space Tourism, 2018, PRRB Publishing, ASIN-B078TLTPVB.

Stakem, Patrick H. Rocket Science-101, PRRB Publishing, ASIN-B079474PD6.

Stakem, Patrick H. In-Space Robotic Repair, 2018, PRRB Publishing, ISBN-9781977066701.

www.ingramcontent.com/pod-product-compliance
Lightning Source LLC
Chambersburg PA
CBHW030512220526
45464CB00006B/2765